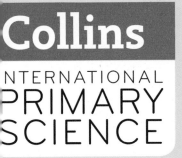

Collins

INTERNATIONAL PRIMARY SCIENCE

Student's Book 5

William Collins' dream of knowledge for all began with the publication of his first book in 1819. A self-educated mill worker, he not only enriched millions of lives, but also founded a flourishing publishing house. Today, staying true to this spirit, Collins books are packed with inspiration, innovation and practical expertise. They place you at the centre of a world of possibility and give you exactly what you need to explore it.

Collins. Freedom to teach.

Published by Collins
An imprint of HarperCollins*Publishers* Ltd.
77–85 Fulham Palace Road
Hammersmith
London
W6 8JB

Browse the complete Collins catalogue at www.collins.co.uk

© HarperCollins*Publishers* Limited 2014

10 9 8 7 6 5

ISBN: 978-0-00-758623-3

The authors assert their moral rights to be identified as the authors of this work.

Contributing authors: Daphne Paizee, Karen Morrison, Tracey Baxter, Sunetra Berry, Pat Dower, Helen Harden, Pauline Hannigan, Anita Loughrey, Emily Miller, Jonathan Miller, Anne Pilling, Pete Robinson.

British Library Cataloguing in Publication Data
A Catalogue record for this publication is available from the British Library.

Commissioned by Elizabeth Catford
Project managed by Karen Williams
Design and production by Ken Vail Graphic Design
Photo research by Emily Hooton

Acknowledgements
The publishers wish to thank the following for permission to reproduce photographs.
Every effort has been made to trace copyright holders and to obtain their permission for the use of copyright materials. The publishers will gladly receive any information enabling them to rectify any error or omission at the first opportunity.
COVER: Paulo Melo / Shutterstock.com
p6 RF Company / Alamy, p10 JOHN BRACEGIRDLE / Alamy, p16 Gib Martinez / Alamy, p30 sciencephotos / Alamy, p42 nguyen khac thanh / Shutterstock.com, p58 wk1003mike / Shutterstock.com, p61 centre Washington Imaging / Alamy, p61 top Outdoor-Archiv / Alamy, p82 Iryna1 / Shutterstock.com, p83 Iryna1 / Shutterstock.com, p84 NASA, p86 NASA / JPL-Caltech

All other photos Shutterstock.

FSC is a non-profit international organisation established to promote the responsible management of the world's forests. Products carrying the FSC label are independently certified to assure consumers that they come from forests that are managed to meet the social, economic and ecological needs of present and future generations, and other controlled sources.

MIX
Paper from
responsible sources
FSC™ C007454

Find out more about HarperCollins and the environment at **www.collins.co.uk/green**

Printed by Grafica Veneta

Contents

Topic 1 Plants

In this topic you will revise what you know about plants. Then you will learn about how plants reproduce and make new plants. You will observe some ways that seeds are dispersed. You will also observe the male and female parts of plants, and investigate what seeds need to grow. You will recognise that flowering plants have a life cycle.

Key words
- roots
- stems
- flowers
- leaves
- absorb

1.1 Plants are living

Plants are living things. We know this because plants grow. As they grow, plants change and make new plants.

Plants look different but they all have **roots**, **stems**, **flowers** and **leaves**.

1   Study and compare these five plants. Name the parts of each plant.

2 Look at the pictures again and revise what you know about plants. Discuss these questions.

- Why do plants need leaves?
- Why do they need roots?
- What do the stems do?
- Why do plants have flowers?

A plant needs soil, water and light in order to grow. The roots of most plants grow in the soil. The roots hold the plant in the soil, and they **absorb** water and other materials from the soil. The leaves of plants use light to make food.

3 List the things that plants need to grow.

Activities

1 Label the picture of the plant in your Workbook on page 1 and describe the job of each part of the plant.

2 Work in groups. Each group grows a plant in the classroom. Observe the plant every day for two weeks. Record your observations in your Workbook on page 2.

3 Predict what would happen if you put your plant in a dark cupboard in your classroom. Would it grow? Explain what would happen and why.

I have learned

- Plants are livings things, so they grow.
- Plants need soil, water and light to grow.

1.2 Plants need energy from light

Animals and plants need food to grow. Animals cannot make their own food, but plants can. So, animals depend on plants for food.

Plants make their own food by using light energy from the Sun through a process called **photosynthesis**. They also use water and air.

In the process of making their food, plants give off **oxygen**, which animals need to breathe and live.

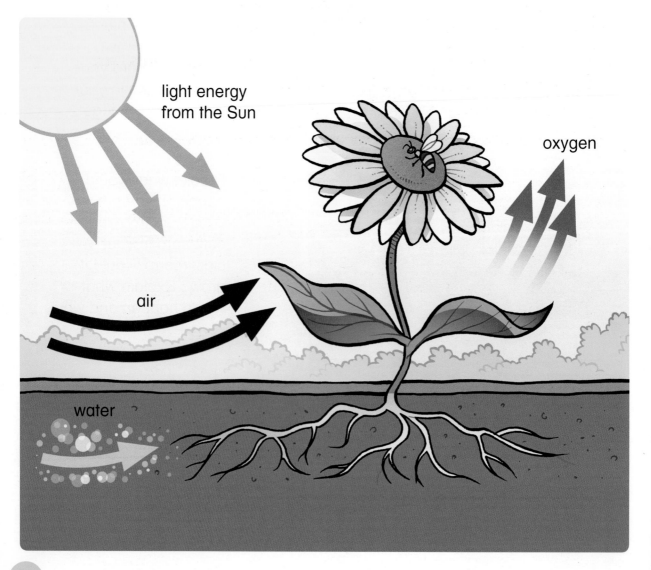

light energy from the Sun

oxygen

air

water

4

Plants use a green substance in their leaves called chlorophyll to make their food. If a plant does not get enough light it will not be able to make food. Eventually, the plant will die.

▲ *Light keeps the leaves of a plant green and healthy.*

▲ *This plant has not had enough light. The leaves cannot make food for the plant to grow.*

1 Work in pairs. Discuss what you need to do to make plants grow well.

2 Look at the leaves in pictures A and B. Discuss the differences. What effect do these differences have on the plant?

3 Work in pairs. Tell your partner how a plant makes its own food.

Activities

1 Write an article for a science magazine. In it, describe two ways in which animals depend on plants.

2 Draw a diagram on page 3 of your Workbook to show what a plant needs to grow.

3 Set up an investigation to investigate if plants need sunlight to grow. Record your investigation in your Workbook on page 4.

I have learned

● Like animals, plants need food to grow. Unlike animals, plants can make their own food.

● Plants use energy from light to make food.

● Plants make food in their leaves.

1.3 Plants can make new plants

Key words

- reproduce
- runner
- seed
- bulb
- corm
- tuber

Plants are living things that breathe, feed and grow. Like all living things, plants also **reproduce**. This means that they can make new plants. Plants can reproduce (make new plants) in different ways.

Some plants grow **runners**. These runners are stems, but they grow along the ground, instead of upwards like other stems. Roots grow from these runners and then new small plants start to form.

1 Look at the pictures of the plants with runners. In what ways are the new plants similar to the big plants? Describe where the runners grow.

Some plants make **seeds** in order to reproduce. These seeds grow into new plants.

2 Look at some seeds from your own environment and discuss where the seeds come from. From which part of the plant do they come?

Some plants reproduce by making **bulbs**, **corms** or **tubers**. Bulbs, corms and tubers all grow under the ground. Roots and shoots grow from them.

A bulb is oval in shape with a point at the top. This point is where the plant shoot starts to grow. A corm is round and flat, and a tuber is a long, swollen root.

a tuber

corms

bulbs

Activities

1 Draw pictures in your Workbook on page 5 to show the different ways in which plants reproduce.

2 Describe three ways in which plants reproduce. Write some sentences on page 6 of your Workbook.

3 Investigate reproduction in plants in your own environment. Make a poster to show the different ways in which the plants reproduce. Draw pictures or take photographs to use on the poster.

I have learned

- Plants are living things that can reproduce (make new plants).

- Some plants grow runners that develop into new plants.

- Some plants reproduce by making bulbs, corms or tubers.

1.4 Flowers help plants to reproduce

Flowers help plants to reproduce because they make seeds. Seeds develop when **pollen** from the male parts of flowers come into contact with the female parts of flowers. Most plants have both male and female parts that help them to reproduce.

1 Study this diagram, which is a **cross-section** of a flower.
Find the male and the female parts of the flower.

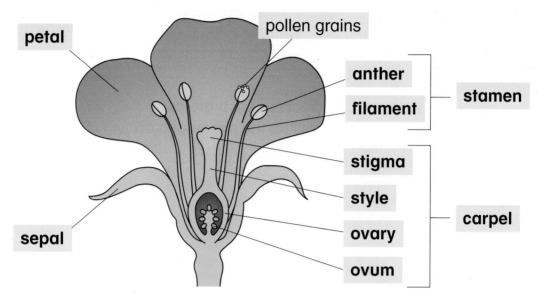

petal

pollen grains

anther

filament

stamen

stigma

style

carpel

ovary

ovum

sepal

▲ *The stamen is the male part of a flower.*
The carpel is the female part of a flower.

2 Why do you think flowers have male and female parts?
Discuss this in groups and report back to the class with your ideas.

Activities

2 Collect four different flowers. Your teacher will make cross-sections of them for you. Use a magnifying glass to see if you can find all the male and female parts.

1 Make a list of the male parts of a flower. Then make a list of the female parts of a flower. Use your lists to label the flower in your Workbook on page 7.

3 Make a detailed drawing of a cross-section of a flower in your Workbook on page 8 and label the parts.

I have learned

● Flowers help plants to reproduce.
● Flowers have male and female parts.

1.5 From flower to seeds

Pollen is made by the anthers, which are at the tops of the stamens. This pollen, from the male parts of a flower, may land on the female part of a flower – the stigma. This is called **pollination**.

A small tube grows from each grain of pollen. It grows down the style and into the ovary. When the pollen and an ovum join together it is called **fertilisation**.

1   Look at this cross-section of a flower. ▼
    Trace the way the pollen moves from the
    male parts of the flower to the female parts.

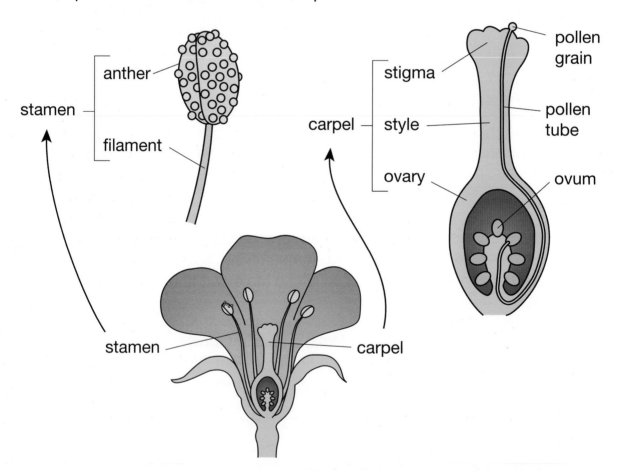

The fertilised part of the flower develops into a **fruit**. The ova become **seeds** and they grow inside the ovary, which becomes a type of protective container for the seeds. We call this a **pod**. The other parts of the flower die.

2 Look at these different kinds of seeds. Where do the seeds grow? What protects them?

dates

dandelion

avocado

poppy

carob

peanuts

broom

tomatoes

Activities

1 Explain to your partner how fertilisation takes place in a flower.

2 Write about what happens when fertilisation takes places. Complete the activity on page 9 of your Workbook.

3 Collect at least six different types of seeds from your environment. Find out which plants the seeds come from. Record your findings in your Workbook on page 10.

I have learned

- Pollen moves from the stamen to the stigma, and then part of the pollen moves down to the ovary.
- Fertilisation takes place in the ovary of the flower.
- Seeds develop inside the ovary.
- The ovary forms a protective container around the seeds.

1.6 Insects and flowers

In order for fertilisation to take place, pollen must first get from the anthers (the male parts) of a flower to the stigma (the female parts) of a flower. This movement of pollen from the anthers to the stigma is called **pollination**. If pollination does not happen, plants cannot produce fruits and seeds.

Pollen grains have little hooks on them, which make them sticky. They stick to the bodies of the insects that land inside the flower. Some of the pollen falls off the insect and on to the stigma, which is also sticky. Sometimes the pollen falls on the stigma of the same flower and sometimes it falls on the stigma of flowers nearby as the insects move from flower to flower.

The colourful petals of a flower attract **insects** like bees. Some flowers produce a sweet substance called **nectar**, which attracts insects. Some insects are attracted by the smell of flowers.

▼▶ *Bees and other insects are good **pollinators**. Birds, slugs and snails can also be pollinators.*

bee

bird

slug

snail

1 Explain how and why flowers attract insects.

2 Read this newspaper headline. ▶ Discuss what it means.

Decline in the bee population threatens food supplies

Activities

1 Explain the difference between fertilisation and pollination.

2 Discuss why pollinators are important to gardeners and farmers.

3 Do some research about bees where you live. Are there fewer bees than before? Why? In what ways has this affected the area?

I have learned

● Pollination takes place when pollen is moved from the anthers (male parts) of flowers to the stigma (female part) of flowers.

● Insects (and birds, slugs and snails) are good pollinators.

● Insects are attracted to flowers by colours and by smells.

1.7 Seeds get around – wind, water and explosion

Plants make seeds in order to reproduce. These seeds need to be **dispersed** or spread around on the soil, so that new plants can grow from the seeds. The quicker and further the seeds are dispersed, the better the chance that new and healthy plants will grow. Seeds can be dispersed by animals, by wind and by water.

1 Look at this picture. ▼
It shows the way that a dandelion spreads its seeds.
Explain what you think happens.

There are many different types of seeds and seed **pods**. They all have features and structures that help them to spread. Some seeds are light, and others are heavy. Some seeds or seed pods have wings, and some have little hooks so that they can stick on to animals.

Some seeds are dispersed by the wind. These seeds are usually light, or have small wings that help them to float in the wind. Other seeds are carried around by water. The seeds of plants that grow near to rivers often float down the river to disperse.

Some seed pods explode to disperse their seeds. When the seeds are fully grown, the pod dries out and then explodes and all the seeds inside the pod are thrown out of the pod by the explosion. The seeds scatter across the ground.

▲ *Some pods explode to disperse seeds! The seeds burst out from the pods and they scatter across the ground. This pod has twisted open and the seeds have burst out.*

2 Name four ways in which seeds are dispersed.

Activities

1 Look at these pictures carefully. Describe each seed and say which way you think it is dispersed.

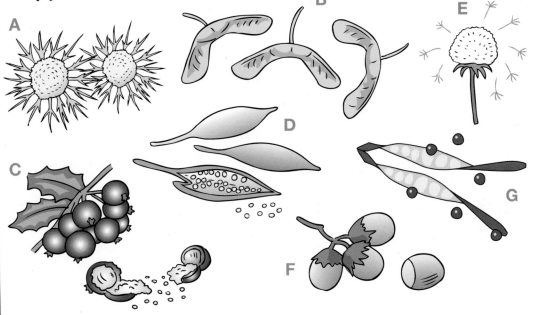

2 Investigate the ways that seeds can be spread. Examine six different types of seeds. Discuss how you can tell which way each seed is dispersed. Record what you find out in your Workbook on page 14.

3 Make a model of a spinner to show the way that some seeds are dispersed. Use your model to demonstrate how seeds can travel.

I have learned

- Seeds from plants can be dispersed (spread around) in different ways.
- Seeds need to be dispersed so that new plants can grow.
- Seeds are different and they have different features which help them to spread.

Key words
- droppings
- burrs

1.8 Seeds get around – animals

Animals can also help seeds to disperse. Some wild animals and birds eat fruit. The seeds of the fruit pass out on to the ground in the **droppings** of these birds and animals.

▲ *Fruit-eating mammals disperse seeds in their droppings.*

▲ *Birds like this bulbul eat fruit and seeds. They disperse seeds through their droppings.*

▲ *Bats are very important in forests. They help to pollinate flowers and they also help to disperse seeds. The seeds spread out all over the forest as the bats make droppings as they fly.*

1 Can you think of any other animals that help to disperse seeds?

▲ *Ants also help to disperse seeds. They collect seeds and take them down into their nests under the ground. They eat the coverings of the seeds and then leave the seeds to grow under the ground.*

Some seeds have small hooks or **burrs** on them. As an animal walks past a plant, the seeds hook on to the animal's fur or hair, and then move away with the animal. People can also help to disperse seeds in the same way. The seeds hook on to shoes and clothes.

2 Look at this picture. What way is this animal helping to disperse seeds?

3 What happens when people throw their apple cores or orange pips on the ground when they have finished eating?

1 Name three animals that help to disperse seeds. In what way does each animal help to disperse the seeds?

2 Observe some animals in your own environment. What do they eat? Predict the way that they could help to disperse seeds. Record your observations and predictions in your Workbook on page 17.

3 Choose your favourite fruit. Find out the way that the seeds of this fruit are dispersed. Make a poster in your Workbook on page 18 with the information that you collect.

I have learned

- Animals, including birds and insects, can disperse seeds.
- Seeds are dispersed through animal droppings.
- Some seeds also have hooks which help them to stick on to animals and move where the animal moves.

1.9 **What do seeds need?**

Not all seeds that fall on the ground will **germinate** or start to grow into a new plant. Millions of seeds are dispersed, but only a few of the seeds germinate and become plants.

Some seeds lie **dormant** (inactive) in the soil for years, waiting for the right conditions to germinate. Many seeds can be dormant for 10 to 15 years. Some seeds can lie dormant for hundreds or even thousands of years!

So why do some seeds germinate while others don't?

1 Work in groups. Discuss what you know about plants and seeds. What do you think seeds need in order to germinate?

▲ *This seed is starting to germinate.*

18

Work in groups to investigate what seeds need to germinate. Follow these steps:

● Prepare four jars. Label the jars A, B, C, and D.

● Put blotting paper or cotton wool in each jar. Put four bean seeds in each jar, between the glass and the blotting paper or cotton wool.

● Put jars A and B in a sunny place in the classroom. Put some water in jar A and keep the cotton wool moist all the time. Do not water jar B.

● Put jars C and D in a cold place, away from sunlight. Put some water in jar C, but do not water jar D.

2 Discuss why you need to use two different jars in each position.

Activities

1 Predict what will happen to the bean seeds in each jar. Explain your predictions and record them in your Workbook (page 19).

2 Observe and record what happens to the bean seeds in your investigation.

3 Over a few weeks, observe the bean seeds that have germinated. Record how the seeds grow into plants. Record the height of the growing plants and comment on what you see. How many leaves are there? Record your observations in your Workbook on page 20.

I have learned

● Seeds need warmth and water to germinate.

1.10 Do seeds need light?

You already know that seeds need water and warmth in order to germinate. You also know that plants need **light** to grow and make their own food. But do seeds need light to germinate?

1 Study this picture carefully. ▼

- Where does the seed germinate? Is it light or dark? Is it warm or cold? Is there water?

- What happens when the seed grows into a small plant? Where are the leaves? What will the plant need to grow now?

Work in groups. Set up your own investigation to find out if seeds need light to germinate. Use bean seeds in labelled jars, as you did in the last investigation.

Activities

1 Discuss what you will do to make this investigation a fair test. Predict what will happen to the seeds. Use page 21 of your Workbook to record this.

2 Record what happens to the bean seeds in your investigation every day for one week.

3 Write a paragraph in which you compare seeds and plants. Does a seed need light in order to germinate? Does a plant need light in order to grow? Why do seeds and plants need different conditions? (Think about the process of photosynthesis.)

I have learned

● Seeds do not need light to germinate.
● Plants need light when they start to grow leaves.

1.11 Growing seeds in variable conditions

Look at these plants. They are all growing in greenhouses.

Greenhouses allow farmers to control the temperature and the amount of water the seeds and plants receive. In a cool climate, a greenhouse can protect seeds and plants from the cold. In a dry climate with very little rain, the seeds and plants can be watered more regularly. Controlling the conditions helps the seeds and plants to grow well.

1   Why are greenhouses useful? What conditions can the farmer control?

2   Do plants grow faster when it is warm or cold?

You already know what seeds need to germinate. Now you are going to investigate how seeds germinate in variable conditions.

You will change the conditions and observe what happens. Some of the things you can change to create variable conditions include:

- the temperature
- the volume of water that you give to each seed
- the amount of light the seed gets.

Activities

1 Work in groups. Discuss what you are going to do to set up an investigation with variable growing conditions. What will you do to make it a fair test?

Ask questions such as:

- What will you use to measure?
- When will you take the measurements?
- Which part of the germinating seed can you measure to find out how much it has grown or not grown?
- How often will you make measurements? Why?

2 Record the measurements you make in your Workbook on page 22.

3 In your group, discuss what you can conclude from the measurements that you have taken. Present your conclusions to your class using your data and evidence.

I have learned

- We need to make accurate and repeated measurements to collect scientific evidence.

1.12 The life cycle of a plant

You have already learned about the different stages in the life of a plant. We can show these stages in different ways.

1 Compare Diagrams A and B, which both show the different stages in the life of a bean plant.

- Which diagram shows that the life of a bean continues in a cycle that does not stop?

- What does the other diagram NOT show you?

- Which diagram do you think gives better information?

Diagram A

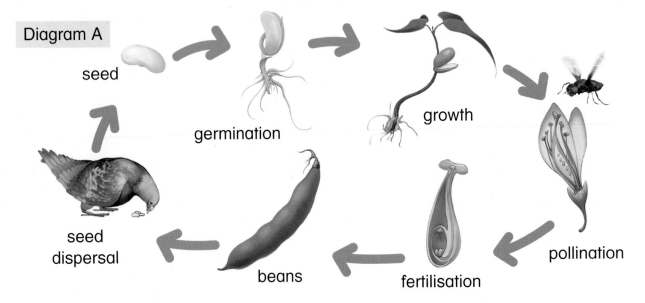

▲ *The life cycle of a bean plant.*

Diagram B

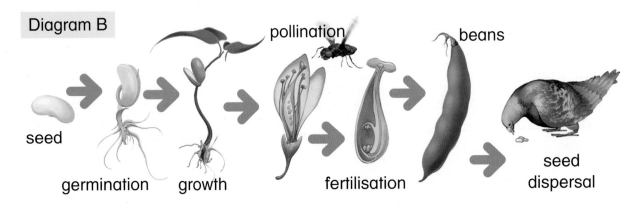

▲ *The life of a bean plant.*

2 Study this **life cycle** and discuss what information it give us about the plant.

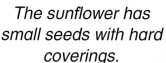

The sunflower has small seeds with hard coverings.

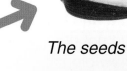

The seeds germinate.

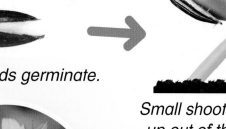

Small shoots come up out of the soil.

sunflower

The plant grows big and flowers start to form.

The flower dies and the seeds are dispersed.

The flowers open and bees pollinate the flowers.

Activities

1 Complete the activity on the life of a plant on page 23 of your Workbook.

2 Look at the stages in the life of a sunflower above. Make your own life cycle diagram to show these stages in your Workbook on page 24.

3 Choose a plant that you know about and make a life cycle poster for it. Describe the way its seeds are dispersed.

I have learned

- Flowering plants have a life cycle.
- We can show the life cycle by means of a diagram.

1.13 Stages in the life cycle

Key words
• life cycle
• pollinated
• ovum (*plural ova*)
• fertilised
• seeds
• germinate

Every stage is important in a **life cycle**. If the life cycle is broken, then it cannot continue. If the flower is not **pollinated**, the **ova** will not be **fertilised** and the plant will not produce new **seeds**. If the seeds do not get any water, they will not germinate and grow into new plants. If the seeds cannot **germinate**, new plants will not grow.

1   Explain, in a very simple way, the life cycle of a plant.

2   Look at this table about the life cycle of a plant. Can you add anything more to the table?

Stage	What happens	What is needed
germination	seeds start to grow under the ground	soil, water, warmth
growth	seeds send up small shoots out of the ground, and the shoots grow into plants	soil, water, warmth, light
reproduction	plants produce flowers with male and female parts	flowers
pollination	pollen from the male anthers comes into contact with the female stigma of the flower	pollen from anthers (male parts of flowers), the stigma (female part), pollinators like insects, birds, water or the wind

Stage		What happens	What is needed
fertilisation		a pollen grain joins up with the ovum and a fruit grows inside the ovary	ovum in the ovary
seed formation		seeds grow inside the fruit	ovary, warmth
seed dispersal		seeds drop from the dying flower and fall on the ground in other places	seeds need to be spread around as much as possible

Activities

1 Discuss what you think might happen if no seeds from one particular type of plant germinate. Could this happen? Why?

2 Draw and label a diagram of the life cycle of a flowering plant. Do this in your Workbook on page 25.

3 Use the final measurements from your plant investigation in Unit 1.11 to produce a class bar chart to show which group's seed grew into the tallest plant.

I have learned

● Each stage in the life cycle of plant is important. If the life cycle is broken, the plant will not reproduce and that type of plant might die out.

Looking back **Topic 1**

In this topic you have learned

- Plants are livings things with different parts that help them to grow and reproduce.

- Plants need soil, water and light to grow.

- Plants are living things that can reproduce (make new plants). They do this by means of runners or seeds.

- Flowers have male and female parts, and these help plants to reproduce.

- Fertilisation takes place when pollen joins with an ovum to form a seed.

- Pollination takes place when pollen is moved from the anthers (male parts) of flowers to the stigma (female part) of flowers.

- Insects are good pollinators. They are attracted to flowers by colours and by smells.

- Seeds from plants are dispersed in different ways: by the wind, by animals, by water and by exploding.

- Seeds need warmth and water to germinate, but not light.

- Flowering plants have a life cycle, which we can show in a diagram.

- If the life cycle of a plant is broken, the plant will not reproduce and that type of plant might die out.

How well do you remember?

1 Draw a picture to show the male and female parts of flowering plants. Label your picture.

2 Explain what plants need in order to grow.

3 Explain the role of this bee in the life cycle of a flowering plant.

4 Draw a life cycle of a flowering plant and label the main stages of the cycle.

Topic 2 States of matter

In this topic you will revise what you know about the three states of matter. You will focus on water and investigate the way that the state of water can change when the temperature changes. You will learn about the processes of evaporation and condensation. You will also find out how rain forms and about the water cycle. You will learn about the boiling point and melting point of water, and about dissolving substances to form solutions.

2.1 Solids, liquids and gases

All the materials (matter) in our environment are **solid, liquid** or **gas**. We call these the three **states of matter**. Each of the states has certain **properties**. We use these properties to decide which state a material is in.

Solids have a fixed shape.

Liquids have a surface but no real shape.

Gases have no surface and no real shape.

Some solids can be poured, but they do not take the shape of their container.

Liquids flow. They can always be poured and they always take the shape of their container.

Gases always take the shape of their container.

1 Look at the pictures of water on page 29 and below. Discuss and describe the state (or states) of the water in each picture.

2 Discuss what properties you think water has in each state.

Water can change from one state to another. The changes happen when there is a change in **temperature**. Think about what happens when ice melts. Ice is water in a solid state. A rise in temperature changes the solid-state water to liquid water. When we boil liquid water, some of the water changes into a gas called water vapour.

Activities

1 Complete the sentences on page 27 in your Workbook to revise what you already know about the states of matter and changes of state.

2 Test the states of matter of some different materials. Complete the observation sheet in your Workbook on page 28.

3 Discuss these questions:
- What change in state takes place when ice melts?
- What change in state takes place when water boils?
- What does the temperature have to do for ice to melt?
- What does the temperature have to do for water to boil?

I have learned

- There are three states of matter: liquid, solid and gas.
- The three states have different properties.
- The three states of water are:
  liquid = water we can pour
  solid = ice
  gas = water vapour.

2.2 Liquid to gas – evaporation

Key words
- evaporate
- rate
- surface area

When water dries up, we say it has **evaporated**. But where does the water go? We know that the water does not disappear. It changes from a liquid state into a gas state and mixes with the air. This process is called evaporation.

These boys are investigating what will happen to the water in the puddle on a hot, sunny day.

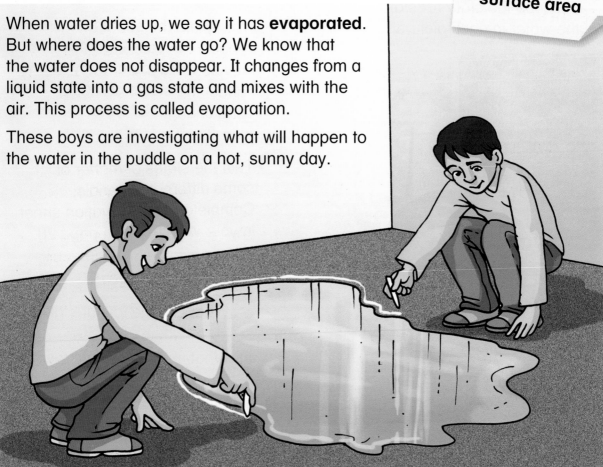

1 Do you think the puddle will stay the same?

2 Where does the water go?

3 Describe what the boys are doing to measure the evaporation.

Water evaporates more quickly in some conditions than in others.

An increase in the temperature can cause water to evaporate more quickly.

The hot air inside a tumble dryer makes the clothes dry more quickly.

An increase in the flow of air around a liquid can increase the **rate** or speed at which it evaporates.

Clothes dry more quickly on a windy day.

The **surface area** of a liquid that comes into contact with the air also affects the rate of evaporation.

Water in a saucer evaporates more quickly than water in a narrow glass, because the surface of water that is in contact with the air is bigger.

4 Name three things that can affect the speed of evaporation. Give an example of each.

Activities

1 Do your own puddle investigation. Create one shallow puddle, one deep puddle and one with a large surface area. Which puddle will dry up first?

2 Work in groups and set up an investigation to find out a way to dry wet clothes quickly. Read the instructions and record your findings in your Workbook on pages 30–31.

3 Set up an investigation to find out the way that the shape of a container can affect how quickly water evaporates. Think about ways to make this a fair test. Record your investigation and findings in your Workbook on pages 32–33.

I have learned

● A liquid can change into a gas, through a process called evaporation.

● Factors such as temperature, air flow and surface area affect the speed of evaporation.

● When a liquid evaporates, the gas mixes with the air.

2.3 Gas to liquid – condensation

When a gas turns into a liquid, we say it **condenses**. You already know that temperature can affect the state of matter. Most gases will turn into liquids if they are cooled down.

1 What happens when steam from a kettle or a hot shower touches a cool tile or mirror or window? In what way is water changing its state?

When water vapour is cooled down, the water vapour changes into liquid. So condensation is the **reverse** of evaporation.

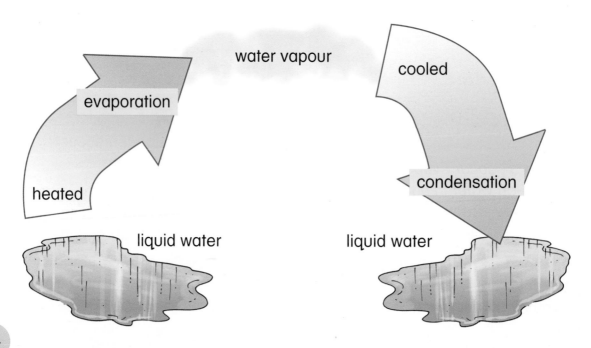

water vapour

cooled

evaporation

heated

condensation

liquid water

liquid water

Look at this picture of an experiment. ▶

warm water

Some hot water was poured into the bottom cup. Then another cup was placed over the bottom cup.

2 What do you think happened after a few minutes? Where? Why? What do you think would happen if we put ice cubes on top of the top cup? Why?

Evaporation and condensation can be useful for collecting water in dry places, such as in a desert. Some animals and plants that live in dry places have adapted to collect water vapour as it condenses.

▲ *Darkling beetles live in the Namib Desert in Africa, which is close to the sea. They stand on top of sand dunes and collect the water vapour that condenses from the air.*

Activities

1 Use a diagram to explain why condensation is the reverse of evaporation.

2 Set up an investigation to find out what happens to ice when it is warmed. Record your results in your Workbook on page 34.

3 Make a poster to show some ways you could collect water if you were in a very dry place. Use page 35 of your Workbook to help you.

I have learned

- Condensation occurs when a gas turns into a liquid.
- Most gases turn into liquids if they are cooled down.
- Condensation is the reverse of evaporation.

2.4 Water vapour in the air

Key words
- water cycle
- reversible

About 71% of the surface of the Earth is covered in water. Water is essential to life on Earth.

Water changes state, but these changes are **reversible**. Water evaporates all the time, but then it condenses again – so the water is never lost.

1 Think about the investigations you have already done and discuss these questions.

- When a puddle of water dries up, what happens to the water? Where does it go?

- When you put wet clothes out to dry, the water evaporates and the clothes get dry. Where does the water go?

When water is heated, it evaporates and changes into a gas (water vapour). This gas mixes with the air and rises. As the gas goes higher, the air gets cooler and cooler. The cooler temperature causes the gas to condense and change back into a liquid. Small drops (droplets) of water form in the clouds, and eventually fall back to the surface of the Earth as rain. Some droplets may freeze, and fall as hail or snow.

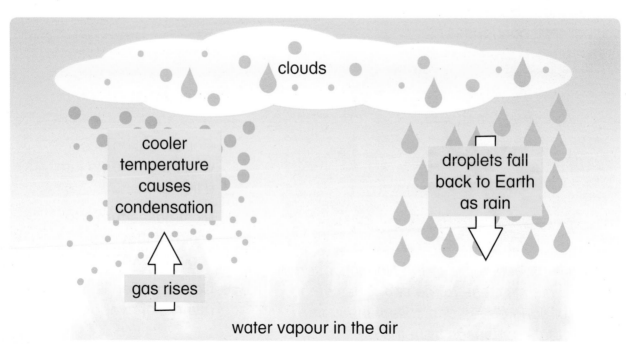

clouds

cooler temperature causes condensation

droplets fall back to Earth as rain

gas rises

water vapour in the air

2 Look at this picture of rain falling on a lake. ▶ Describe how water changes from one state to another state in this area.

Water moves around in a cycle on Earth. This is called the **water cycle**. The same water goes round in a cycle forever! Life on Earth depends on this water cycle. If water did not change state, there would be no life on Earth.

3 Now look at this picture. ▶ Discuss how snow can be part of the water cycle.

Activities

1 Find out some interesting facts about water in its different states on Earth and produce a slide presentation, or storyboard, to present to the rest of the class.

2 Make a list of all the places on Earth where water can evaporate.

3 Explain to your partner where rain comes from. Use pictures or draw your own diagrams.

I have learned

● Water changes state, but it is never lost.

● Water evaporates and rises up into the air as a gas (water vapour).

● Water vapour turns back into a liquid when it is cooled.

● Water droplets form in the clouds and fall to the Earth as rain.

2.5 The water cycle

Water moves around the Earth in a cycle. We call this the **water cycle.**

1   Study the diagram carefully and read the captions. ▼
    Then describe what happens in the water cycle.

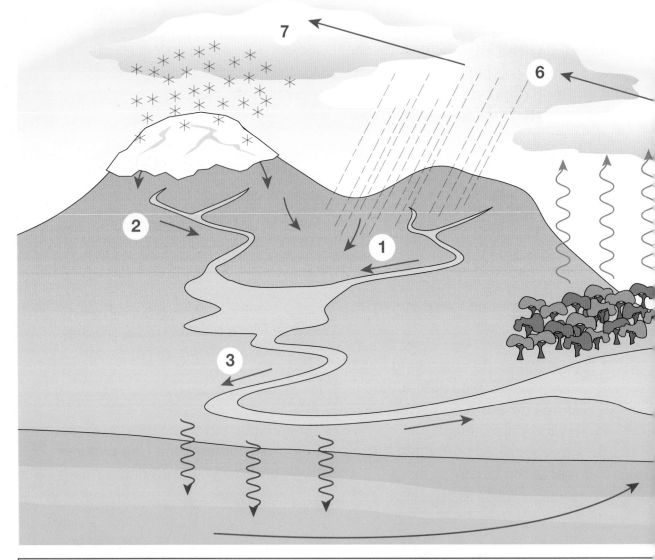

1   Rain falls and collects in streams and rivers. This is called 'run off'.

2   Snow falls on high mountains. Some of the snow melts and runs off into rivers.

3   Rivers flow into lakes and into the sea. Some water sinks into the ground.

4   Water on the surface of the sea evaporates as a result of heat from the Sun.

5   Water vapour rises up until it reaches the cooler air high up in the sky.

38

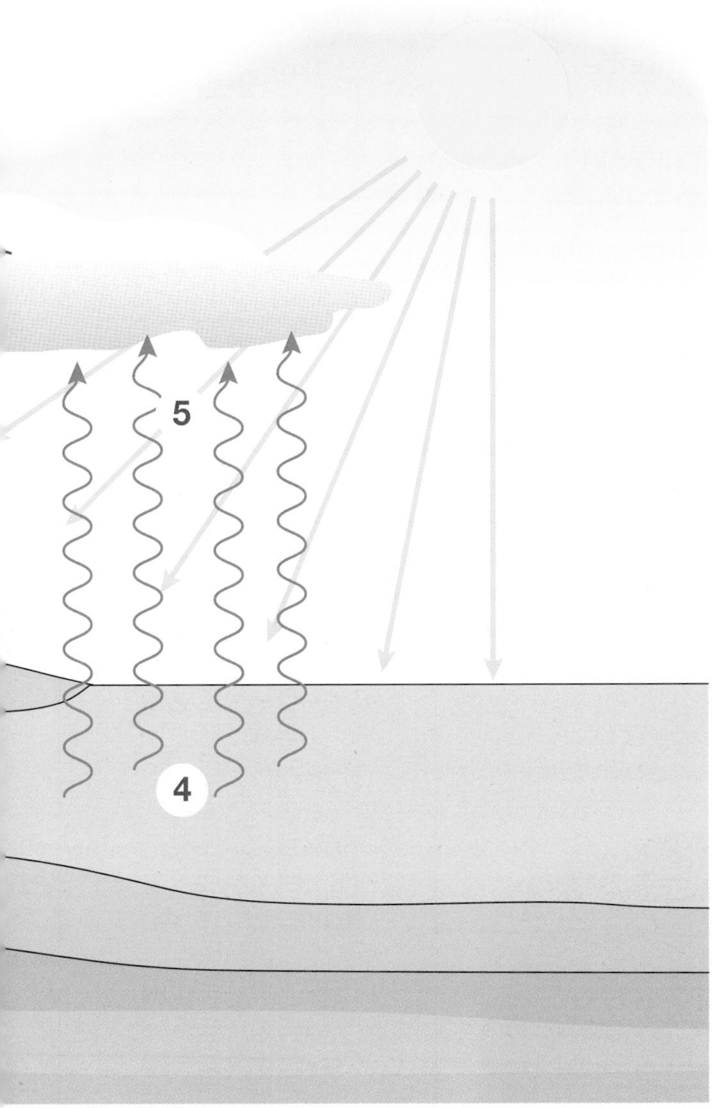

Activities

1 Explain why the water cycle is known as a 'cycle'. What other cycles do you know?

2 Work in groups and model the water cycle. Follow the instructions in your Workbook on page 36.

3 Write a story. Imagine that you are a droplet of water. Create a story to describe your journey around the water cycle.

5

4

6 Some of the water vapour condenses into droplets in the clouds.

7 When the water droplets are big enough, they fall as rain. Sometimes the droplets freeze and fall as snow or hail.

I have learned

- Water moves around the Earth in a water cycle.

- The processes of evaporation and condensation are essential parts of this cycle.

- The Sun plays an important part in the water cycle.

2.6 Boiling and freezing

Key words
- temperature
- degrees
- thermometer
- Celsius
- boiling point
- melting point
- freezing point

Solid, liquid and gas are the three states of matter. Matter can change state though processes like evaporation and condensation. They can also change state through melting, freezing and boiling. A change in **temperature** is usually needed for a change of state to happen.

Temperature is the measure of how hot or cold something is. We measure temperature in **degrees** using a **thermometer**. We use the **Celsius** scale (°C) for measuring temperature.

1   Use a thermometer to measure the air temperature at different places inside and outside your classroom. What do you notice?

Water boils when it reaches a temperature of 100 °C. This is the **boiling point** of water. When water boils, bubbles form in the water and steam leaves the surface of the water. Steam is evaporating water vapour.

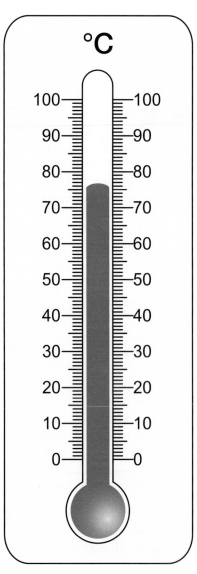

2   When water boils, it evaporates more quickly. Why do you think this happens?

Melting happens when a solid changes into a liquid. Ice, which is water in a solid state, melts at a temperature of 0 °C and turns into a liquid. This is the **melting point** of ice. This is also the temperature at which water usually starts to freeze. We call this the **freezing point** of water.

Activities

1 Explain the terms 'boiling point' of water and 'melting point' of ice. What is the difference?

2 Place a glass of ice in your classroom. Measure the temperature inside the glass and the air temperature inside the classroom. Predict what will happen to the ice. Complete the activity on page 38 of your Workbook.

3 Investigate the way that the temperature of ice changes when it is heated. Record the results in your Workbook on pages 39–40.

I have learned

- We use a thermometer to measure temperature.
- The boiling point of water is 100 °C.
- The freezing point of water is 0 °C.
- The melting point of ice is 0 °C.

2.7 What happens to substances dissolved in water?

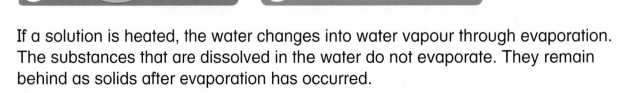

Key words
- dissolve
- solution
- solute
- solvent

Some substances **dissolve** in liquids to make **solutions**. The substance that dissolves is called the **solute** and the liquid in which it dissolves is called the **solvent**. Water is a common solvent.

1 Predict which of these substances will dissolve in water:

> sugar   soap powder   pepper   salt   clay

2 Look at these pictures carefully. Describe the process that is shown in the pictures. What has happened to the solute?

If a solution is heated, the water changes into water vapour through evaporation. The substances that are dissolved in the water do not evaporate. They remain behind as solids after evaporation has occurred.

Seawater contains a lot of salt. The salt is dissolved in the water. In many countries evaporation is used to remove the water from seawater. Then the salt that is left behind is collected.

▶ *Natural salt evaporation ponds at Khanh Hoa, Vietnam.*

Activities

1 Set up an investigation to test if sugar dissolves. Collect the items that you need and record your findings in your Workbook on page 41.

3 Find out more about ways in which salt is made in different places around the world. Does salt production happen in your country?

2 Discuss the photograph of the salt pond on this page. Write your answers to these questions:

● What can you see around the piles of salt?

● What is the man raking?

● What process is taking place?

● In what ways do you think the weather conditions affect this process?

I have learned

● Some substances can dissolve in water to make solutions.

● When water evaporates from a solution, the substances that are dissolved in the water are left behind as solids.

2.8 Getting the solid out of a solution

Key words
- solvent
- solute
- saturated solution
- crystal

A solution is a mixture of a liquid **solvent** and a **solute**. We can make a **saturated solution** by making the solvent hold the greatest possible amount of a solute, until it can hold no more. To do this we usually have to heat the solution.

1 Discuss what you could do to make a saturated solution using salt and water.

If the water in a solution evaporates, a solid is left behind. If the solution is saturated, **crystals** may form as a result of the evaporation. The temperature and the rate at which the solution evaporates both affect the size of the crystals that are formed. For example, if the water evaporates quickly, the crystals will probably be quite small.

2 Why does a solution need to be saturated in order for crystals to form?

Salt crystal

Sugar crystals

3 Look carefully at these pictures. The students are making crystals by using the process of evaporation. Discuss the way that their method works.

Activities

1 Follow the instructions your teacher gives you to make some sugar crystal candy.

2 Follow the instructions your teacher gives you to make some salt crystals.

3 Investigate ways of making crystals. Record your investigation in your Workbook on page 42.

I have learned

● When a liquid evaporates from a solution, a solid is left behind.

● Saturated solutions are needed in order for crystals to form.

● Factors such as temperature and rate of evaporation can affect crystal formation.

Looking back Topic 2

In this topic you have learned

- There are three states of matter: liquid, solid and gas.
- The three states have different properties.
- The three states of water are: liquid = water we can pour; solid = ice; gas = water vapour.
- A liquid can change into a gas, through a process called evaporation.
- Temperature, air flow and surface area affect the speed of evaporation.
- When a liquid evaporates, the gas mixes with the air.
- Condensation is the reverse of evaporation. Condensation occurs when a gas turns into a liquid.
- Water moves around the Earth in a water cycle, so it is never lost.
- Water evaporates and rises up into the air as a gas (water vapour) and then turns back into a liquid when it is cooled. Water droplets form in the clouds and fall to the Earth as rain.
- We use a thermometer to measure temperature. The boiling point of water is 100 °C. The freezing point of water is 0 °C. The melting point of ice is 0 °C.
- We can get the solid back from a solution through evaporation.

How well do you remember?

1 Look at these pictures. ▶
   Name the state of water in each picture. Then say which process you could use to change the state of water shown into another state.

2 Explain why evaporation is the reverse of condensation.

3 What is the difference between the boiling point of water and the melting point of ice?

4 Give an example of a way in which evaporation is useful in our daily lives.

Topic 3 Light

In this topic you will learn about light. First, you will find out about the way that shadows are formed, and how their size, length and direction can change. Then you will learn to measure the intensity of light, exploring materials that let light through and materials that do not let light through. You will begin to learn about the role of light in how we see things and you will explore the way that light beams can change direction when they are reflected.

3.1 Making shadows

Key words
- shadow
- light
- source
- beam

Shadows are formed when **light** travelling from a **source** shines on something but cannot pass through it. Some materials can block the light.

Light travels in straight **beams** from a light source to the objects we see. If there is no light source, it is dark and we cannot see things.

1 Look at these pictures. What do they tell us about the way light travels?

2 What happens to a light beam when it hits a tree trunk? Describe what will be seen behind the tree.

3 Look at these photographs carefully. ▼ ▶ Say what you know about the shadows in the photographs.

- What is a shadow?
- What causes a shadow?
- On which surface do you see the shadow?
- What time of day do you think it is? Why?

4 Predict what would happen if light beams could bend around the edges of objects such as tree trunks.

Activities

1 Your teacher will make the classroom as dark as possible. Shine a flashlight on different objects and see what kinds of shadows you can create. Use page 43 in your Workbook to record what you did.

 Go outside the classroom on a sunny day and stand in an open area. Mark the place where you stand and your shadow. Repeat this a few hours later. Use page 44 in your Workbook to record the shape and size of your shadows each time.

3 Discuss what factors can affect the size of a shadow. Use information from your observations and the investigations that you carried out.

I have learned

- Some materials block light because the light cannot travel through them; this causes shadows.
- Light travels in straight lines from a light source.

3.2 Shadows outside

The Sun is the main light source outside. As the Earth spins around each day, the Sun appears to move across the sky. The Sun rises in the east each day and sets in the west. You will learn more about this in the next topic.

The position of the Sun in the sky affects the length, direction and shape of shadows cast on Earth. Shadows outside change their shape, length and direction during the day.

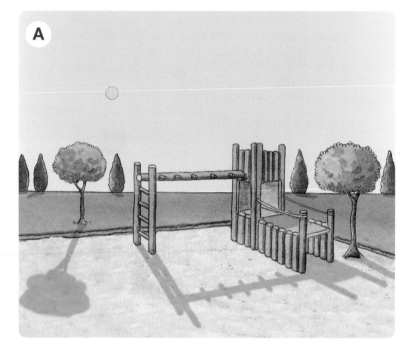

1   Study pictures A and B. Look at the position of the Sun in the sky and describe the shadows. What are the differences between the two pictures?

2   Name three ways in which shadows outside change during the day.

Think about the activity that you did earlier, when you observed your own shadow outside. You recorded changes in the shape, length and direction of your shadow between your two observations.

3   Why did your shadow change shape, length and direction?

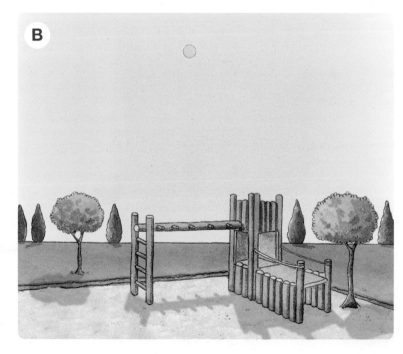

Look at these pictures. ▼
Read the labels.

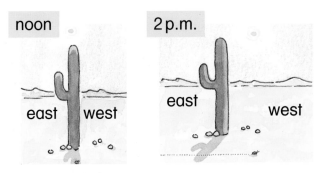

1 Investigate the way that shadows change during the day. Model this using a flashlight to represent the Sun, and a toy or classroom object to make the shadows. Record your investigation in your Workbook on page 45.

2 Take two objects outside. Observe and measure their shadows at least six times during the course of a day. Record your results in your Workbook on pages 46–47.

3 Explain the way the movement of the Earth makes shadows change throughout the day. Use a model of the Earth and the Sun to do this.

4 What are the differences between the shadows?

5 Explain why the shadows are different.

I have learned

- The Earth's movement causes the Sun to appear to move in the sky during the day.

- The shape, length and direction of shadows on Earth change during the day because the Sun's position in the sky changes.

3.3 Changing the size of a shadow

Shadows are bigger when the object is closer to the **light source**. Light spreads out from the light source, so a near object blocks out more of the light and makes a bigger shadow.

Study this picture carefully. The girls are creating a shadow. They are shining the light source (flashlight) on to a cardboard puppet. The puppet is blocking the light from the source and is forming a shadow on the wall.

1 Predict what will happen to the shadow if the girls keep the light source in the same place, but move the puppet closer to it.

2 Predict what will happen to the shadow if the girls move the light source further away from the puppet.

Look at this picture. ▼

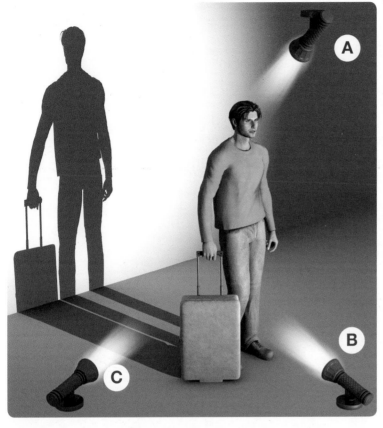

Kwesi is making a scary movie. In this scene the main character is standing with the light in front of him. Kwesi has asked the lighting engineer to position a spotlight so that the character will **cast** a big, long shadow on to the wall behind him.

3 Where would be the best position for the lighting engineer to place the spotlight – at A, B or C? Why?

Activities

1 Look at the picture of the investigation on page 52 and collect everything you will need to copy it. Write down your predictions on page 48 of your Workbook.

2 Do the investigation and test your predictions. What did you do to make sure the test was fair?

3 Use a ruler and record your measurements on page 49 of your Workbook. Write a conclusion.

I have learned

● The size of shadows can change.

● Shadows are bigger when the object is closer to the light source, because the object blocks out more of the light from the light source.

● Shadows are smaller when an object is further away from the light source.

3.4 Recording shadows

People have made clocks that work with shadows for a very long time. We call these shadow clocks **sundials**. The earliest sundials come from Ancient Egypt and Babylon (which is now part of Iraq). Some of them are over 5000 years old.

Sundials can be used to tell the time because the shadows that are made on Earth are predictable. The shadow made by the same object, in the same place, on the same day of the year and at the same time of day will always be the same. This is because the Earth and the Sun are always in the same positions at a particular time on a particular day. You will learn more about this later.

Look at the lines and numbers on these sundials. ▼

1 In what way are these different from the lines and numbers on a normal watch or clock?

2 Explain how a sundial works.

3 Could you use a sundial at night?

A group of students observed an object's shadow outside at different times of the day. They drew a diagram to record their observations.

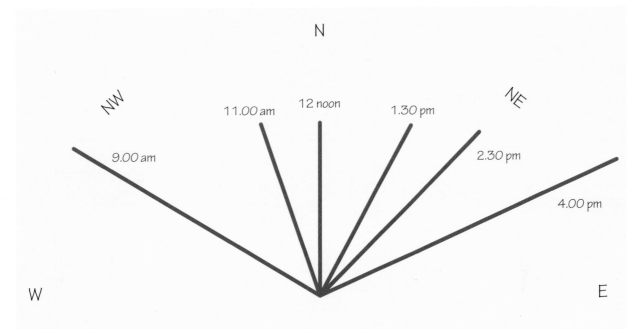

4 Discuss what the diagram shows.

Activities

1 Explain why people used sundials thousands of years ago.

2 Set up an investigation to measure the direction and length of an object's shadow at different times of the day. Draw a diagram in your Workbook on page 50 to record the results. Draw conclusions about the longest and shortest shadows.

3 Build your own sundial. Line it up with north and use it to tell the time.

I have learned

- The shape, length and direction of shadows on Earth change during the day because the Sun's position in the sky changes.
- Sundials use shadows to show the time.

3.5 Materials and light

Light shines from a light source. It travels in straight lines from the source. Some materials can block light as it travels, and that is how shadows are formed. But other materials allow light to pass through them as it travels.

Materials can be **transparent, translucent** or **opaque**.

▲ Opaque materials block all the light. They do not allow any light to pass through them. We cannot see through these materials.

▲ Transparent materials allow light to pass through them. We can see through these materials clearly.

▲ Translucent materials let some of the light through. We can see through these materials a little, but we cannot see clearly.

1 Explain what the difference is between opaque, translucent and transparent materials.

2 Think about a sheet of paper in an exercise book. Is it opaque or translucent? Explain your answer.

3 Look at this helmet. Which parts are transparent and which parts are opaque? Why?

▼ *Sunglasses are translucent. They have a special coating on them which protects your eyes by blocking out harmful light from the Sun.*

Activities

1 Write down the names of three objects in your home that are made from opaque materials and three objects made from transparent materials. Say what the materials are and what the objects are used for.

2 Investigate if materials are opaque or translucent. Record your results in your Workbook on page 52.

3 Some foods are packed in transparent containers and some are packed in opaque containers. Find examples of each and explain to the rest of the class why you think they are packed using these materials. Record your findings in your Workbook on page 53.

I have learned

- Opaque materials do not let light through.
- Transparent materials let a lot of light through.
- Translucent materials let some light through.

3.6 Playing with light and materials

People use opaque and translucent materials to make shadows in some types of theatre.

1 Talk about the way in which this shadow puppet theatre has been set up. What is the light source? Where is it?

2 Explain how the puppet theatre makes use of opaque materials.

3 The screen that is used is translucent. What would the audience see if the screen was opaque? What would they see if the screen was transparent?

▲ *Shadow puppets in a theatre in Thailand.*

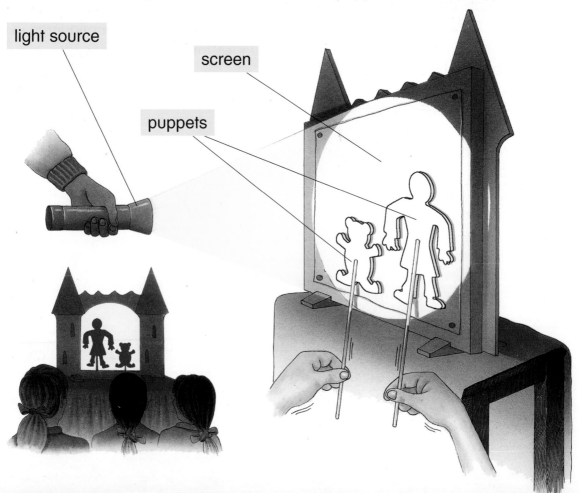

light source

screen

puppets

Shadows are not always the same. Sometimes shadows can be dark and black, like in the shadow puppet theatre. These shadows have sharp edges and they are easy to see. But shadows can also be in colour and they can be **fuzzy**. They will be fuzzy if some of the light is blocked, but not all the light.

4   What do you think will happen if you put an opaque coloured material in front of a light source? Will you get a shadow? What will the shadow look like?

5   What do you think will happen if you put a translucent coloured material in front of a light source? Will you get a shadow? Will you see colours?

Activities

1   Design a stained-glass window. Follow the instructions in your Workbook on page 54.

2   Set up an investigation to find out about the different types of shadows that can be formed. Record your findings in your Workbook on page 55.

3   Explain how a shadow puppet theatre uses the behaviour of light. Draw a diagram, label the diagram and write a short explanation.

I have learned

● Shadows are not always black and clear.

● Shadows can be in colour.

3.7 Can we measure light?

Have you ever thought about what people mean when they say 'It's *bright* outside' or 'It's a *dull* day'? They are talking about the **intensity** of the light outside. The intensity is how bright (strong) the light is.

1 Look around the classroom. Then look outside. Can you describe the light? Is there a difference? Why?

Our eyes can tell the difference between bright light and dull light. To measure light accurately, we can use a **light meter**. The unit that we use to measure the brightness (or illuminance) of light is the **lux**. The abbreviation is lx. This measures the intensity of the light as it appears to our eyes. It does not measure the energy of the light source.

The table gives some typical lux measurements. The exact values can vary as they depend on reflections from surrounding surfaces and colours that are reflected.

2 Use information in the table. Predict what lux measurements you would get today:
- inside your classroom
- on a sports field outside.

Time of day or place	Lux (lx) measurement
direct sunlight	32 000–130 000
full daylight (not direct sunlight)	10 000–25 000
shady place on a cloudy day	11 500
overcast day	1000
sunrise or sunset	400
inside an office	400
full moon on a clear night	1.00
starlight	0.0011

3 Discuss why light meters can be useful. When might people want to measure the intensity of light? Why?

Activities

1 These lux measurements all describe the intensity of light. Put them in order from the brightest to the dullest:

1.00   11 000   1750   0.011 800

2 Use a light meter to measure the intensity of light in different places around your school. Record your findings in your Workbook on page 57.

3 Find out about other units that are used to measure light. For example, find out what a lumen and a candela are.

I have learned

- We can measure the intensity of light with a light meter.
- The intensity of light can be measured in units called lux.

3.8 When do we need to measure light intensity?

Light is very important in photography. A photographer needs to know what the intensity of the light is around the object or person that they are photographing. They need to know this so that they can set up their **camera** to take good clear photographs.

▲ *Using a light meter*

Some photographers use separate **light meters**. They take several readings around the object or person they want to photograph. They use the measurements to set the aperture of the camera lens. The aperture is the space or opening through which the light passes into the camera lens.

Photographers also set the speed at which they take the photograph. This is the speed at which the shutter on the camera opens and closes. If the light intensity is quite high, they will take the photograph using a fast shutter speed. If there is very little light, they will need to use a slower shutter speed in order to get enough light to take the photograph.

Most modern cameras have built-in light meters. The camera measures the intensity of the light and adjusts the settings automatically, so all the photographer needs to do is to press the button to take the photograph!

▲ *This camera has a built-in light meter, which measures the light automatically before taking a photograph.*

There are other uses for light meters too. For example, you already know that plants need light to make their own food. The intensity needs to be about 1200–2000 lux for the process to take place. If a farmer grows plants indoors, it is useful to measure the intensity of the light and to increase it if necessary, to make sure the plants grow well.

There are other uses for light meters too. For example, you already know that plants need light to make their own food. The intensity needs to be about 1200–2000 lux for the process to take place. If a farmer grows plants indoors, it is useful to measure the intensity of the light and to increase it if necessary, to make sure the plants grow well.

1 Think about people who design buildings (architects). Why do you think they might use light meters?

Activities

1 Look at some light meter readings on page 59 of your Workbooks and complete the activity.

2 Explain why photographers use light meters.

3 Take some photographs using different apertures and speeds, and compare the photographs. Then discuss why the light was important in each photograph.

I have learned

- Light meters measure the intensity of light.
- Light meters are useful in photography.

3.9 Seeing light

Light travels from a **light source**. Objects that give out their own light are called light sources. The Sun, a flashlight, a light bulb that is switched on, a flame from a candle and the stars in the sky are all light sources.

We can see objects because light from a light source shines on the objects, **bounces** (**reflects**) off them and enters our eyes. If there is no light, we cannot see at all.

▲ *This boy can see what is written in the book because light travels from the light source, is reflected off the book and then enters his eyes.*

Opaque materials block light that reaches our eyes. This can cause shadows. If a material is translucent it blocks some of the light that reaches our eyes. Light enters our eyes and falls on the inside parts of our eyes that are very sensitive to light. Translucent materials can therefore protect our eyes.

1 Work with a partner. Look carefully at your partner's eyes but don't touch them. Can you see the point where the light enters the eyes? It looks like a black spot, but it is really a hole.

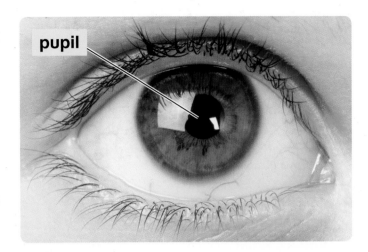

▶ *The black spot (called the* pupil) *is where light enters the eye.*

You already know that sunglasses are translucent and have a special coating on them which protects your eyes by blocking out harmful light from the Sun.

2 Explain how sunglasses protect your eyes.

Activities

1 Draw a picture of your partner's eye. Discuss why you think there is a hole in the middle of the eye.

2 Complete the diagrams and sentences on page 60 of your Workbook.

3 Draw two light sources on page 61 of your Workbook and explain why they are light sources.

I have learned

● We can see light sources because light from the source enters our eyes.

3.10 Reflecting light

We can see objects because light from a light source shines on the objects, **bounces** off them and then enters our eyes. When light bounces off a surface, we say that the light is **reflected** by the surface.

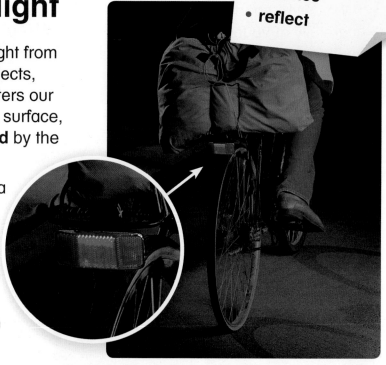

1 This bicycle reflector is not a light source, so how do we see it? Where does the light come from and how does it reach our eyes?

2 Read this story. Discuss what you have learned from the story.

In the stock room …

I can't see anything.

That's because we have no source of light in here.

Dalia switches on the flashlight.

We can see the flashlight now.

But we can see everything else too, even the things that aren't light sources.

Dalia has a bright idea.

That must mean that all the other things reflect light because when the light source is off, we don't see them …

… and when the light source is on, we do see them.

The Sun is a light source and so we can see it clearly. The Moon is not a light source, but we can often see the Moon quite clearly in the sky. We can see the Moon when it reflects the light of the Sun.

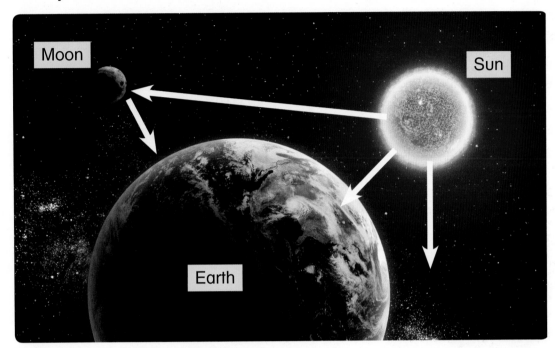

Moon

Sun

Earth

Activities

1 Complete the diagrams in your Workbook on page 63 to show that you understand the way that light is reflected.

2 Create a comic strip story like the one on page 66 to help explain to younger children about light sources and reflected light.

3 Find out more about the way that the Moon reflects light from the Sun. Why can we sometimes only see part of the Moon? Can we ever see both the Sun and the Moon in the sky at the same time? Draw diagrams to explain your answers.

I have learned

- Light can be reflected from some surfaces.
- When light reflected from an object enters our eyes, we can see the object.

3.11 Reflecting and absorbing light

Some materials reflect light better than other materials. Smooth surfaces, like mirrors, and shiny surfaces, such as metal or foil, reflect light so well that we can see images of ourselves on them.

1 Why do you think we can see the reflections so clearly in this lake?

Other materials, like paper and plastic, reflect light well enough for us to see the object, but not well enough for us to be able to see our image on the surface of the object. These materials **absorb** some of the light that shines on them, and the rays of light also reflect off and **scatter** from some surfaces.

2 Look at these diagrams and discuss what they show about the way that light is reflected from two different types of surfaces.

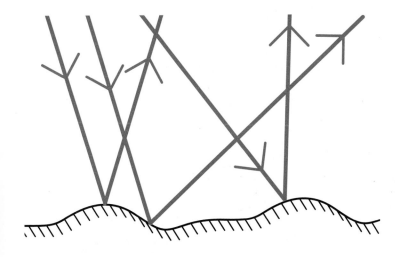

▲ *Reflections on a rough surface*

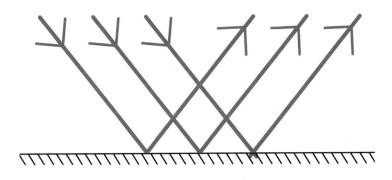

▲ *Reflections on a smooth surface*

Activities

1 Give two examples of surfaces that reflect light very well. Say why they reflect light so well.

2 Set up an investigation to find out how well different materials do or don't reflect light. Record your investigation in your Workbook on page 64.

3 Demonstrate to the class the way that one material reflects light better than another material. Make a poster or draw a diagram to show why the reflections are different.

I have learned

- The surfaces of some materials reflect light better than others.
- Mirrors, shiny surfaces and still water are good reflectors of light.

3.12 Changing the direction of light

Key words
- angle
- periscope

Light travels in straight lines, and the rays or beams of light do not bend. So is it possible to see things that are behind us or above us? What can we do to see things that are around a corner or over a wall?

We can use mirrors to change the direction of beams of light. When light hits a mirror, it is reflected off the mirror and changes direction. The light reflects off the mirror at a certain **angle**.

To understand the way that light is reflected off a mirror, look at this picture of a snooker player hitting a ball against a surface. The ball bounces off the side of the table at the same angle as it hit it. Light behaves in the same way.

This picture shows a girl using a **periscope** to look at some children playing on the other side of the high wall.

1   Trace the direction of the light from the children to the eyes of the viewer. What do you think is inside the periscope?

2 Discuss what you could do to look at the back of your own head. Explain a way that you could use mirrors to do this. Try this out and then make a sketch and draw the direction of the beams of light. Start your arrows at the light source and finish at your eyes.

Activities

1 Use a flashlight and a mirror and play with the light. Make the light move in different directions.

2 Investigate how you can get light to move around a corner. Use a mirror or mirrors and a flashlight. Then draw a diagram in your Workbook on page 66 to show what you did.

3 Build a game. Use black paper and wooden blocks to make a race track. Try to get light from a flashlight to shine all along the track by using mirrors.

I have learned

- A beam of light can change direction when it is reflected from some surfaces.

71

Looking back Topic 3

In this topic you have learned

- Light travels in straight lines from a light source.

- Shadows are formed when materials block light from a light source.

- Shadows are bigger when the object is closer to the light source, because the object blocks out more light from the source.

- The shape, length and direction of shadows on Earth change during the day because the Sun's position in the sky changes.

- Materials can be opaque (do not let light through), transparent (let a lot of light through) or translucent (let some light through).

- Shadows are not always black and clear; they can be in colour.

- We can measure the intensity of light with a light meter, in units called lux.

- We can see because light enters our eyes, either directly from a light source or because it is reflected off a surface.

- The surfaces of some materials reflect light better than others.

- A beam of light can change direction when it is reflected from some surfaces.

How well do you remember?

1   Name two light sources.

2   Draw a picture to show the way that shadows change during the day. Label your picture and write two or three sentences to explain your drawing.

3   Draw a simple diagram to explain how we see things. Your diagram should include a light source and a drawing of the eye. Label your diagram and use arrows to show the direction of light.

4   What is the name of the process in which light changes direction? Give one example of a way in which this can be useful.

Topic **4** The Earth and beyond

In this topic you will make a model in order to explore the ways that the Earth moves. You will learn that the Earth spins on its own axis and that it orbits the Sun. Then you will research the lives and discoveries of some of the scientists who have studied the Solar System and the stars.

4.1 Where does the Sun go at night?

Scientists now know that the Moon moves around the Earth and that the Earth moves around the Sun, but for a long time people thought that the Sun moved around the Earth. This is because it appeared to travel across the sky from the east to the west every day.

▲ *In this picture, a photograph of the Sun was taken every 30 minutes and the images were put together to show the way in which the Sun rises and appears to move across the sky from east to west.*

Today, we still say that the Sun 'rises' at a time that we call sunrise. And we say that the Sun 'sets' at a time that we call sunset, even though we know that it is really the Earth that moves and not the Sun. The Sun rises in the east and sets in the west.

1 Why does the Sun appear to move across the sky?

The Earth is shaped like a **sphere**, it moves around the Sun once a year and it also spins as it moves, in an *anti-clockwise* direction (from west to east). As the Earth spins around and parts of it come to face the Sun, the Sun seems to rise up in the sky. As the Earth spins away from the Sun, the Sun seems to sink down below the **horizon**.

2 What does a sphere look like?

3 What direction would the Earth be spinning in if it moved from east to west?

We can **model** the Earth spinning to show the reason why we get night and day. When it is daytime where we live, it is night-time on the opposite side of the Earth.

4 Where does the Sun go at night?

5 If it is daytime where you live, is your part of the Earth facing the Sun or facing away from the Sun?

▲ *This photograph shows the Earth as seen from space. The countries that are facing the Sun have daylight.*
The other countries are in darkness, so there it is night-time.

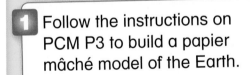

Activities

1 Follow the instructions on PCM P3 to build a papier mâché model of the Earth.

2 In your groups, use your model to demonstrate that the Earth is spinning and the Sun stays still, so that when one part of the Earth is facing the Sun it is daytime and on the other side it is night-time.

3 Use your model to explore this idea: *Sunrise (and sunset) will not occur at exactly the same time in all parts of our country.*

I have learned

- The Earth moves around the Sun.
- The Sun rises in the east (sunrise) and sets in the west (sunset).
- We can use a model to show the way the Earth's spin causes day and night on Earth.

4.2 The Earth rotates on its axis

Think about the way in which you modelled the movement of the Earth in Unit 4.1. What did you do to make the Earth spin or turn around? You probably put a stick through the middle of your model of the Earth.

The globe in the picture has a stick through its middle too. But of course there is no stick through the centre of the real Earth!

The Earth spins or **rotates** on its **axis**. This is an imaginary line that runs from the North Pole through the centre of the Earth to the South Pole.

▶ *A globe is a three-dimensional model of the Earth.*

It takes the Earth 24 hours (one day) to complete one rotation, in other words, to turn around completely on its axis once.

1   How many complete turns on its axis will the Earth make in one week?

2   What would happen if the Earth did not rotate on its axis?

If you modelled the Earth carefully, you will have placed your stick at an **angle** because the axis of the Earth is tilted at an angle. Look at the picture of the globe and the diagram below. Can you see which way the Earth is tilted?

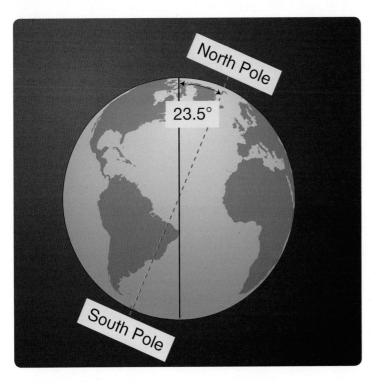

North Pole

23.5°

South Pole

▲ *The Earth is tilted at an angle.*

3 What angle is the Earth tilted at?

4 What would happen if the Earth took 12 or 36 hours to turn on its axis?

Activities

1 Read the statements on page 67 of your Workbook and say if they are true or false.

2 Answer the questions in your Workbook on page 68.

3 Work in groups. You know that the Earth moves around the Sun. Think about the way the Earth is tilted on its axis. What effect do you think this angle has on life on Earth? What would happen if the Earth were not tilted at an angle? Use the model you have made or a globe to help you.

I have learned

- The Earth turns around or rotates on its axis.
- The Earth's axis is tilted at an angle.
- The Earth takes 24 hours to make one complete rotation.
- We have day and night on Earth because of the Earth's rotation on its own axis.

4.3 The Earth's orbit

The Earth rotates on its own axis and, as a result, we have day and night on Earth. At the same time as it is rotating on its axis, the Earth is also moving around the Sun.

The Earth moves in an **orbit** around the Sun. An orbit is a path that an object takes around something else. It takes one year (365 days, or 12 months) for the Earth to orbit the Sun. As a result of these movements and the fact that the Earth is tilted, we have day and night and also different temperatures on Earth.

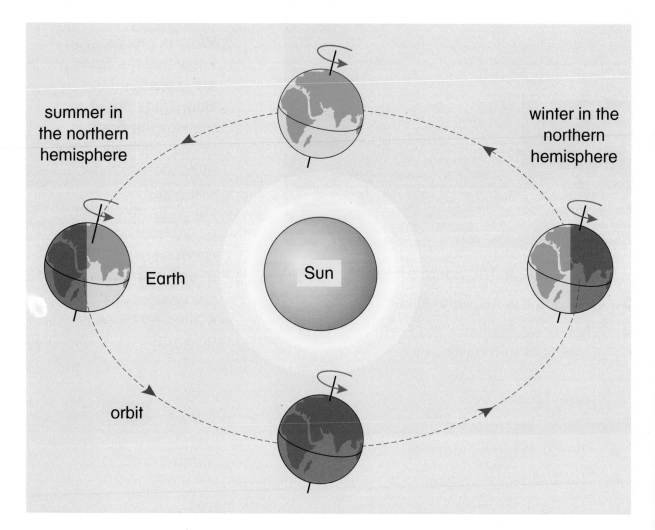

▲ *The Earth's orbit around the Sun is like an oval rather than like a circle.*

1 Describe two ways in which the Earth moves.

2 Study the diagram and think about the way the Earth rotates as it moves around the Sun. What effect does this have on the Earth?

The tilt of the Earth affects the hours of sunlight we get. In the parts of the Earth tilted away from the Sun, the daytime will be shorter. In the parts tilted towards the Sun the daytime will be longer.

6:00 pm summer

6:00 pm winter

Activities

1 Observe where the Sun is in the sky at a particular time of day over the next few weeks. Record your findings in your Workbook on page 69.

2 Work in groups. Use your model of the Earth and a flashlight to explore how the tilt of the Earth affects the hours of daylight in a particular country.

3 Use the data on page 70 of your Workbook to plot a graph to show changes in the number of the hours of sunlight for two different places.

I have learned

- The Earth moves in an orbit around the Sun.

- It takes one year (365 days) for the Earth to complete one orbit around the Sun.

4.4 The Solar System

The Earth is not the only **planet** to orbit the Sun. There are other planets and there are also **moons**, which orbit around the planets. The Sun, the planets and their moons are all part of the **Solar System**.

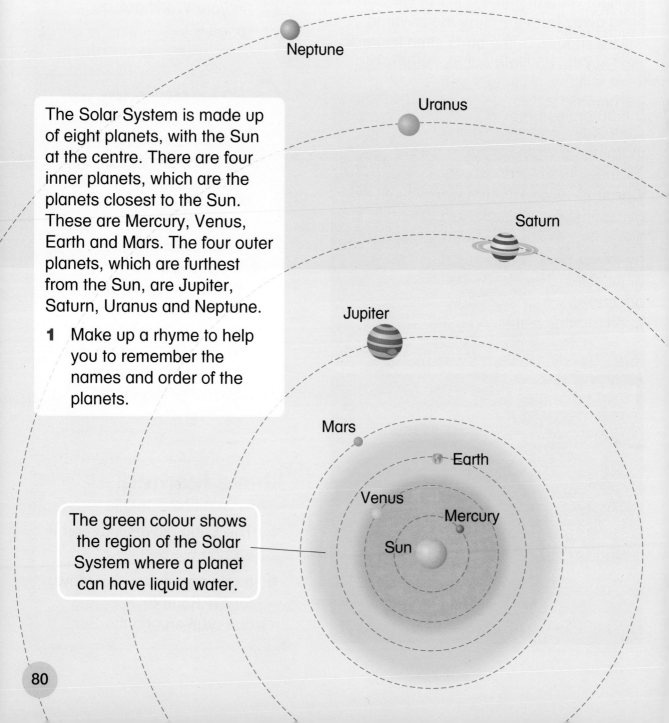

Neptune

Uranus

Saturn

Jupiter

The Solar System is made up of eight planets, with the Sun at the centre. There are four inner planets, which are the planets closest to the Sun. These are Mercury, Venus, Earth and Mars. The four outer planets, which are furthest from the Sun, are Jupiter, Saturn, Uranus and Neptune.

1 Make up a rhyme to help you to remember the names and order of the planets.

Mars

Earth

Venus

Mercury

Sun

The green colour shows the region of the Solar System where a planet can have liquid water.

Earth is at the right distance from the Sun to be warm enough for life. It also has liquid water on its surface, and life needs water. Another important thing about Earth is that it has an atmosphere. This is the mixture of gases surrounding Earth. Without these gases, the heat from the Sun would be lost from Earth and it would be much, much colder at night-time.

2 Explain the factors that make Earth a suitable planet for plants and animals to live.

People have been studying and learning about the Solar System for many thousands of years. **Constellations** of stars were identified by ancient civilisations, and the seasons were measured and predicted by farmers who observed the changing night skies. They knew that certain constellations or stars could only be seen at particular times of the year.

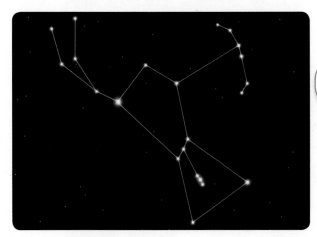

3 In ancient times, in what way could observing the night skies help a farmer to know when to plant their crops?

Activities

1 Create a classroom display of the Solar System to show the Sun and the eight planets which orbit it. Most of the planets have their own moons. Include any that you find out about on your display.

2 Find out about a constellation, such as Ursa Major or Orion, and draw a map of it. Research any ancient stories connected with it.

3 Do some research to find out about one of the planets in the Solar System. Create a poster and present your findings to the class.

I have learned

- The Earth is the only planet in the Solar System which has perfect conditions for life.

- There are eight planets in the Solar System.

- The constellations and stars we can see in the night sky change throughout the year.

4.5 Early astronomers and discoveries

Key words
- astronomy
- Universe
- theory
- telescope
- gravity

Over thousands of years, people gathered more and more knowledge about the Solar System. This knowledge developed into the science called **astronomy**.

In 150 AD, the famous astronomer Ptolemy believed that the Sun moved around the Earth. By 500 AD, the Indian astronomer Aryabhata had discovered that the Earth spun on its own axis. But he still thought that the Earth was the centre of the **Universe**.

In the early 11th century, the Arab scientist Ibn al-Haytham made observations and calculations that questioned the idea of the Earth-centred Universe. By the 15th century, scholars like Al-Qushji had found more evidence that the Earth moved around in the Sun, and in 1514 the Polish astronomer Nicolaus Copernicus also put forward this **theory**. But many people did not want to believe him.

1 What is an astronomer?

▲ *The astronomer Nicolaus Copernicus (1473–1543)*

A few years later, the Italian scientist Galileo built a more advanced **telescope**, based on an earlier one he had seen. Using his telescope, he discovered Jupiter's four largest moons. He observed the moons moving around Jupiter. This evidence supported the ideas of Al-Qushji and Copernicus.

▲ *Jupiter and four of its moons.*

▶ *Galileo Galilei (1564–1642)*

Finally, in the 17th century, the mathematician Johannes Kepler confirmed the motion of the planets around the Sun and Issac Newton's laws of motion and discovery of **gravity** provided a scientific explanation for the work done by Al-Qushji and Copernicus.

2 Name five scientists who have contributed to the development of astronomy.

3 What has been the biggest change in scientific thinking about the Solar System in the last 600 years?

Activities

1 Find out how many moons Jupiter has and explain why Galileo could not see them all with his telescope.

2 Compile a timeline of important discoveries in astronomy in your Workbook on page 73.

3 Do your own research about the life and discoveries of a scientist who investigated the Solar System and the stars.

I have learned

● Astronomers are scientists who study the Universe and all of the objects within it.

● Scientists have proved that the Earth moves around the Sun.

83

4.6 Space exploration today

Key words
- satellite
- space station
- astronaut

Today, we have very sophisticated equipment to help us explore and understand the Solar System. Scientists in many different countries work together to do research, and share their ideas and information.

The USSR sent the first object into space in 1957. This was a **satellite** called Sputnik 1. The first human to go up into space was the **astronaut** Yuri Gagarin, in 1961. And, on 20 July 1969, the American Neil Armstrong was the first human to walk on the Moon.

There are thousands of human-made satellites orbiting the Earth. Some of them help us to predict the weather by collecting images of the Earth from space, while others are communication satellites.

1   What is a satellite?

2   What is the difference between a natural and an artificial satellite?

3   Who was the first person to go up into space?

The Hubble telescope is a very powerful telescope that was launched into space in 1990. It has taken thousands of photographs of deep space. One of its most important discoveries has been to give an estimated age for the Universe of about 13–14 billion years.

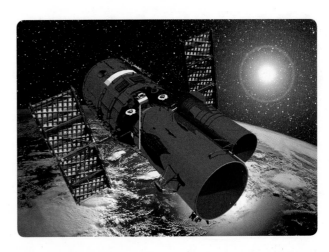

▶ *The Hubble telescope has been up in space since 1990.*

The International **Space Station** (ISS) was launched in 1998. It completes 16 orbits of the Earth every day, and can be seen from Earth at night. Scientists living on it carry out many experiments, both inside and outside the ISS. One of the things that the scientists are researching is the way living in space affects the growth of plants and the health of the human body.

4 Do you think it is important to find out more about the Solar System? Debate this question in class.

Activities

1 Find out why scientists want to know how the human body is affected by living in space for long periods of time.

2 Do your own research to answer the questions in your Workbook on page 76.

3 Do your own research on modern space programmes. You could research the work done by the Indian, Chinese, American or Russian space programmes. Make a report detailing their major discoveries.

I have learned

● Satellites orbiting the Earth send back important information about the weather and climate.

● Modern scientists all over the world are still learning about the Solar System.

● Scientists live, work and carry out lots of experiments on the ISS.

4.7 Into the future

Key words
- asteroid
- comet

Scientists would like, one day, to send humans to explore Mars. In recent years, a number of different countries have sent expeditions to Mars.

The Mars Rover, Curiosity, was sent there by NASA in 2011. It is an unmanned craft which is exploring the surface of the planet and is transmitting data back to Earth. Its missions include trying to find evidence of water and of life on Mars.

▲ *An illustration of the Mars Rover, Curiosity.*

1 Suggest why scientists have sent a rover-type craft to Mars, rather than sending people.

2 Why are scientists interested in exploring Mars?

In the future, it might be possible to build a manned space station on Mars. Having a local source of water would be essential for humans to be able to live and work there.

▲ *The comet Hale-Bopp passing close to the Earth in 1997.*

In the Solar System there are orbiting lumps of rock called **asteroids**. There are also orbiting **comets**.

Asteroids are irregular shaped rocks which orbit the Sun. They vary in size from slightly smaller than a planet, to the size of a grain of sand.

Comets have an icy core. They have distinctive gas 'tails' which can sometimes be seen without a telescope. Early astronomers observed many comets that passed close to Earth. Halley's Comet has been known about since prehistoric times, but it was recognised as being a returning object by Edmond Halley in 1705. He correctly realised that the same comet is visible from Earth every 75–76 years. Halley's Comet will next appear in 2061.

Scientists believe that, in the future, we will be able to mine asteroids and comets for precious metals, minerals, gases and even water!

3 Why, in the future, might it be useful to be able to mine metal and minerals from asteroids?

Activities

1 Imagine that you are in charge of controlling the Mars Rover. Describe how you would use it to find out more about the planet.

2 Research the history of the exploration of Mars, from the invention of the telescope to the modern day.

3 Compile a list of reasons for and against mining asteroids and comets. Consider things such as how much it might cost and how practical it would be.

I have learned

● A comet has an icy core and a gas 'tail'.

● An asteroid is a lump of rock orbiting the Sun.

Looking back Topic 4

- The Sun appears to move across the sky from east to west, but it is really the Earth that moves.

- The Earth rotates on its axis and takes 24 hours to make one complete rotation. This causes day and night on Earth.

- The Earth's axis is tilted at an angle.

- It takes one year (365 days) for the Earth to complete one orbit around the Sun.

- There are other planets that orbit the Sun in the Solar System.

- Modern scientists all over the world are still learning about the Solar System.

How well do you remember?

1 Explain why the Sun appears to move across the sky.

2 Write down six facts that you have learned about the Solar System in this topic.

3 Draw a diagram to show why we have day and night on Earth. Label your diagram.

4 Look at this picture. Why were instruments similar to this important in the development of our knowledge of the Solar System?

5 Write a paragraph about one scientist who has contributed to our knowledge of the Solar System.

Glossary

absorb	To take in something or to soak it up.
anther	The part of the stamen of a plant that contains the pollen.
angle	The amount of turn between two lines that meet at a point, usually measured in degrees.
astronaut	Someone who goes into space.
asteroid	A small rocky object that orbits the Sun.
astronomy	The area of science that deals with space and the Universe.
axis	An imaginary line around which something rotates.
beam	A ray or shaft of light that can be seen, for example from a torch or spotlight.
boiling point	The temperature at which a liquid changes into a gas or vapour.
bulb	An underground organ of a plant that stores food; new plants can be grown from bulbs.
bounce	When something moves quickly up, back or away from a surface after hitting it.
burr	The rough and sometimes prickly outer covering of a seed, which can stick on to an animal's fur.
camera	A device used for taking and storing photographs.
carpel	The female part of a flower.
cast	To cause a light or shadow to fall on a surface.
Celsius	A scale for measuring temperature, in which water freezes at 0 degrees (0°C) and boils at 100 degrees (100°C).
comet	A small object with an icy core that orbits the Sun; when a comet passes near to the Sun it has 'tail' of gas.
condense	To change from a gas or vapour into a liquid.
constellation	A group of stars which form a shape in the sky and that has been given a name, for example Ursa Major.
corm	An underground organ of a plant that stores food, similar to a bulb; new plants can be grown from corms.
cross-section	The cut surface of something that has been cut in half, or a picture to show it.
crystal	A piece of a mineral that has formed naturally into a regular shape.
degree	A unit for measuring temperature or angle, often written as ° after a number.
disperse	To spread over a wide area.
dissolve	To mix something into a liquid to form a solution.

dormant	When something is still alive but does not grow until the conditions are suitable.
droppings	The undigested remains of animals' food.
equator	An imaginary line that circles the Earth at an equal distance from the North and South poles.
evaporate	To change gradually from a liquid into a gas.
fertilisation	In a plant, when a male structure (pollen) combines with a female structure (ovum) to form a seed.
filament	The thin part of the stamen of a plant, which supports the anther.
flower	The colourful part of a plant which contains the organs from which the fruit or seeds develop.
freezing point	The temperature at which a liquid changes into a solid.
fruit	The sweet fleshy part of a plant that contains the seed or seeds; some fruits can be eaten.
fuzzy	When describing light, something is fuzzy if the edges are blurred or it cannot be seen clearly.
gas	A substance that is not liquid or solid, for example the air; gases have no fixed shape.
germinate	When a seed starts to grow.
gravity	The force that makes things fall when you drop them.
horizon	The line where the Earth's surface and the sky appear to meet.
insect	A small animal with six legs, a hard outer shell and usually wings.
intensity	The measurable amount of a property, for example light.
leaves	The plural of leaf, which is the flat green part on the end of a twig or branch of a tree or other plant.
life cycle	The series of changes a plant (or an animal) goes through during its life.
light	The brightness that comes from things such as the Sun, stars and lamps.
light meter	A device for measuring the intensity of light.
light source	A device or natural feature that is a source of light.
liquid	A substance that is not a solid or a gas, for example water; liquids can be poured.
lux	A standard unit for measuring light intensity.
melting point	The temperature at which a solid changes into a liquid.
model	An imitation of something, used as an illustration of the real thing.
moon	A large natural object in orbit around a planet.
nectar	A sweet liquid produced by flowers; it is attractive to insects.

opaque	A property of a material which means you cannot see through it at all.
orbit	The curved path followed by an object going around the Sun or a planet.
ovary	The part of the carpel of a flower that contains the ova; it develops into a seed case after the ova have been fertilised.
ovum (plural **ova**)	The structure in the ovary of a flower which develops into a seed when it is fertilised.
oxygen	A colourless gas that is part of the air.
periscope	A device, containing a set of mirrors, which allows the observer to see things that would otherwise be out of sight.
petals	The brightly coloured outer parts of a flower.
photosynthesis	The process by which plants make their food using sunlight, carbon dioxide and water.
planet	A round object in space which orbits the Sun or another star and is lit by light from it.
pod	A long narrow seed container that grows on plants such as peas or beans.
pollen	A yellow powdery substance found in the flower of a plant.
pollination	The transfer of pollen to the female organs of a plant.
pollinators	Insects and other animals which transfer pollen from the male parts of a flower to the female parts.
properties	The qualities or features of a particular object or thing.
rate	The speed or frequency at which something happens.
reflect	To bounce something back without absorbing it, for example light or sound.
reproduce	To produce young animals or new plants.
reverse	The opposite of something.
reversible	A change that can take place in either direction.
roots	The parts of a plant that grow under the ground.
rotate	To turn about an axis.
runner	A stem that grows from the base of a plant to produce new plants.
satellite	An object in orbit around a planet.
saturated solution	A solution in which no more solute can be dissolved.
scatter	To disperse something in various random directions.
seeds	The small, hard parts of a plant from which new plants can grow.
sepals	The green, leaf-like, outer parts which enclose the petals and other parts of a flower.

shadow	The dark shape made on a surface when an object prevents light from reaching it.
Solar System	The Sun and all the planets, comets and asteroids that orbit round it.
solid	A substance that is not a liquid or gas, for example wood; solids have a fixed shape.
solute	The substance that is mixed with a liquid to form a solution.
solution	A liquid in which another substance has been dissolved.
solvent	A liquid that can dissolve other substances.
source	The place that something comes from.
space station	An artificial satellite that is used as a base for scientists in space.
sphere	A completely round solid object, for example a ball.
stamen	The male part of a flower, which produces pollen.
states of matter	The different ways in which substances can exist - solid, liquid and gas.
stem	The long thin central part of a plant above the ground that carries the leaves and flowers.
stigma	The part of the carpel of a flower on which pollen falls.
style	The long slender structure in a flower which connects the stigma and ovary.
sundial	A device that shows the time by casting a shadow.
surface area	The total area of all the surfaces of an object.
telescope	A device used for looking at distant objects, for example the planets.
temperature	The measure of how hot or cold something is.
thermometer	An instrument used to measure temperature.
theory	An idea or set of ideas used to explain something.
translucent	A property of a material which allows light to pass through it, but you cannot see through it.
transparent	If something is transparent, you can see through it clearly.
tuber	A part of the stem of a plant that is underground; for example, a potato is a tuber; new plants can be grown from tubers.
Universe	Everything that exists.
water cycle	The continuous process in which water evaporates from the sea, condenses to form clouds in the sky, falls as rain, makes its way back to the sea ... and so on.